THE STORY OF TOWER BRIDGE

KEN WILSON

SERIES EDITOR · SHEILA DUNN

LANTHORN · LONDON

First published 1985

ISBN 0 947987 10 X

Lanthorn Publishing Limited
23 Greenwich High Road
London SE 10 8JL

When HRH The Prince of Wales, later King Edward VII, officially opened Tower Bridge on 30 June 1894 he was presented with a silver loving cup to commemorate the occasion. A medal was struck to mark what was regarded as "a structural triumph of this age of steel". Thousands had gathered to see "The Wonder Bridge" and a great cheer rose from the assembly as the Prince turned the specially provided lever to raise the two 100 feet long bascule bridge sections.

Alexandra, The Princess of Wales, who accompanied her husband, received a jewelled and enamelled brooch in the shape of the mark of the Bridge House Estates Committee, which had financed the project as part of an ancient charter dating back to the Middle Ages.

A procession of ships sailed through what was already being called "The Gateway of London. There had been criticism of the design in its Gothic mantle but in spite of the cost rising to a 'staggering' £1,184,000, Tower Bridge was hailed as a masterpiece of Victorian engineering.

The building of the bridge was the culmination of circumstances which had became acute during the previous hundred years but which really went back to the Middle Ages and before that to the Roman occupation of Britain.

Some time in the latter half of the first century AD, a Roman engineer constructed a bridge across the Thames to serve the newly established capital city of Londinium. It probably occupied a site close by the present London Bridge and the level of the bridge made it impossible for larger ships to pass further upstream. The port became fixed downstream of the crossing.

By the twelfth century, another bridge had been built to replace the first and this firmly closed off the river to all but the smallest vessels, confining the Port of London to the Pool - the stretch of river between London Bridge and the Tower - where it remained until the docks were built, commencing with St Katharine's in 1804.

THE BRIDGE IN THE MIDDLE AGES

The origins of the Bridge House Trust, linked with the Corporation of the City of London, began with the early wooden bridges in the 12th century.

During the reign of Edward I, a hundred years later, a toll of one farthing for pedestrians and a penny for horse and rider was introduced. Receipts were to be used for maintaining the 'pontage for London' - to ensure road crossings of the river.

Control of the bridge was vested in the Common Council of the City, which paid Wardens to administer the accounts, grant leases and deal with the routine supervision. The records of the first Lord Mayor of London, Fitz Alwyn, are still preserved at Bridge House and include early accounts, leases and other legal documents.

Today the Bridge House Estates Committee of the City of London Corporation is responsible to the Court of Common Council for the care and upkeep of all the City's bridges - Blackfriars, Southwark, London and Tower - without cost to the public by way of toll. With that responsibility has been included the task of ensuring that not only foot and vehicular traffic is kept moving but also the river's shipping.

The house mark of the Bridge House Trust has existed in some form since the beginning of the 18th century. It is not known what the design signifies. It has been suggested that it derived from the initials 'R. A' of Sir Robert Acherley, Lord Mayor in 1511. Others have associated it with heraldry. Another claim is that it was a merchant's emblem.

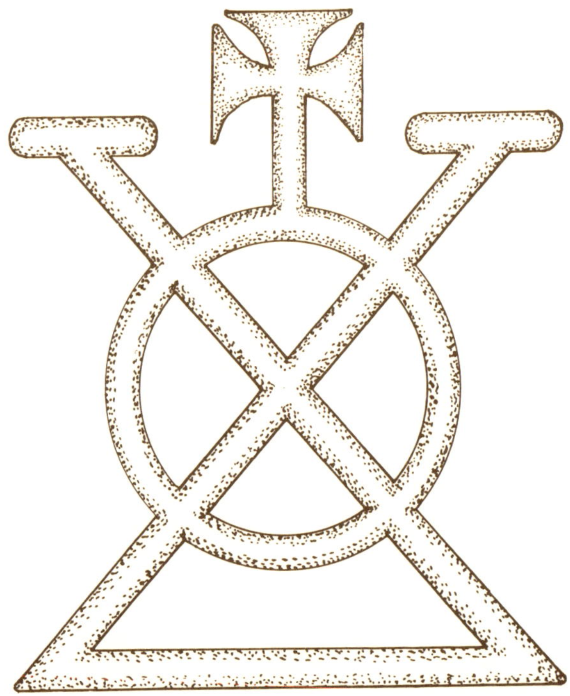

Since it was founded, the Trust's headquarters have occupied a site on the south bank of the river, close by the bridge, approached from what is now Tooley Street. By the beginning of the 19th century a substantial building with an imposing clock tower housed the administration which included residential and office space, a counting house and a large hall which was also used as a court and meeting room where the business of Southwark Council was transacted.

London Bridge was the only road crossing of the Thames until Westminster Bridge was built in 1738. In an attempt to cater for increasing traffic, the buildings had all been removed and, to improve shipping problems, the Great Arch, giving wider and high passage space, was constructed at the centre.

Blackfriars and Southwark Bridges were added in the early years of the 19th century but these were all to the west and there was tremendous commercial and residential development downriver, which involved major detours to make the crossing.

OLD LONDON BRIDGE AT THE BEGINNING OF THE 19th CENTURY

As the City expanded eastwards, the demand for another crossing on that side of London Bridge increased but until sufficient docks were constructed to make it possible to move the port downriver and until engineering could solve the problems involved, this remained an intractable and frustrating difficulty for which no practical solution could be found.

In 1876 the City Architect, Horace Jones, was asked to prepare a report on the feasibility of such a project and examine possible alternatives.

The stretch of river from London Bridge to the Tower was known as the Pool of London. Since the middle ages, this had been the City's port and with the expansion of the empire it became the largest port in the world. Progressively the facilities became incapable of dealing with the volume of traffic and there were intolerable delays, often of months, in handling cargoes. The building of the docks during the first half of the century eased the problem and dramatically reduced the amount of ships using the Pool.

The City Architect's survey showed that the average number of ships passing west of St Katharine's Dock over six days was 144 and the height of masts ranged between 40 and 95 feet. Any bridge built in the suggested position would have to be designed to deal with both shipping and road traffic.

The location of the lock entrance to St Katharine's Dock dictated the line of a new bridge as immediately east of the Tower. The river is about 850 feet wide at this point.

Conventional ideas were considered first. A bridge of the necessary height and crossing the river in a single span was obviously preferable from the shipping point of view but this was at the expense of foot and vehicular passengers who would all have to make the journey up to the height of about 120 feet. Telford and Brunel had solved larger problems but a bridge of this scale within the City limits was not regarded with favour. It would have required enormous amounts of land on both banks for the approaches and would have dominated the Tower.

Reducing the spans, by introducing piers, would have brought corresponding savings in cost and a less daunting structure but this too suffered from the same fundamental defects and sacrificed the major users to the shipping interests.

A scheme which favoured land traffic over river demands was the Duplex system. With circular passing places for ships there would be no stoppage of crossing traffic at all. Only one sliding portion of each would have to be moved, leaving the other in use. A substantial amount of the shipping was barge and small vessels, which would not need to pass through the gates. As might have been expected, the objections to this scheme came from the owners of tall ships.

A rolling bridge was an attempted compromise but the possibility of mechanical failure and other likely risks ruled this out as a serious contender.

Sir Horace Jones received his knighthood for his services as Architect and Surveyor to the City Corporation, the appointment he held for twenty three years. In this office he designed a number of major buildings, including the Billingsgate, Leadenhall and Smithfield Markets. He added the Library and Museum to the Guildhall and the memorial in Fleet Street which marks the original site of Temple Bar.

The son of a prosperous solicitor, he studied architecture and spent some years in Italy and Greece to further his classical knowledge. After setting up in private practice at the age of twenty seven he quickly established a profitable business, designing numerous large office blocks in the capital.

From 1882 to 84 he was President of the Royal Institute of British Architects.

The City bridges came under his control and he prepared a scheme for widening London Bridge but this did not obtain Government approval.

Whilst preparing his report of 1876, on the need for another road crossing, he became convinced that the best solution would be a suspension bridge, supported from two tower piers, with a central opening section 200 feet wide for the larger ships.

As no one else had suggested this, he produced his own design.

In spite of his early studies in classical architecture, Sir Horace had become more and more interested in the revived Gothic style, made fashionable by Barry and Pugin for the new Houses of Parliament, and it was in this idiom that he designed the stone cladding for the twin towers.

The opening sections were originally intended to be lifted by steel cables running from the tops of the towers, through the steel arch, down to the meeting centres of the gates. There were serious engineering flaws in this idea. The wear and tear on the cables would be enormous, the cables would cause obstruction and the precise movement at the junction would be difficult to control.

The overall appearance of the main arch aroused criticism and there were objections that foot passengers could not cross when the gates were raised.

Sir Horace's basic concept was accepted as the best solution. It was a compromise between the conflicting forms of traffic but in general it was more acceptable than any of the alternatives.

The Consulting Engineer appointed was Sir John Wolfe-Barry. He too had a distinguished career, with an impressive list of major achievements to his name. He also had risen to the Presidency of his chosen profession, the Institution of Civil Engineers.

Sir John was the younger son of the same architect whose work so inspired Horace Jones, Sir Charles Barry. The name Wolfe was assumed by Royal Licence in 1899.

Among a number of consultancies, he was retained by the Bridge House Estates Committee.

In collaboration with Wolfe-Barry, Sir Horace Jones produced a revised design. To meet the pedestrian demand for continuous crossing facilities, two wide footbridges at high level spanned between the towers. These also served as bracing to strengthen the whole structure and pacified the objectors to the appearance of the great archway.

Work commenced on site on 22 April 1886. The estimated cost of the bridge and all necessary land acquisition was £750,000 which was well within the capacity of Bridge House Estates to finance and there would be no toll required from road or river users.

The two main towers were built on foundations 204 feet by 100 feet. The river bed was 34 feet deep and a further 12 feet of clay was excavated to find a firm base. Using iron caissons - metal boxing, without top or bottom - sunk into the river bed, it was possible to isolate working areas from the fast-flowing river. After the water inside the caissons had been pumped out, work could proceed in a normal manner. Wooden posts were driven into the clay around the working areas on which temporary scaffolding was laid to provide access and loading space for materials and hoisting gear.

The weight of the bridge is carried by a steel and iron framework - the stone exterior is just a thin cladding. Nearly 12,000 tons of metal was used in the structure.

To give the necessary clearance of 140 feet above the waterline, beneath the footbridges, the two main towers have rectangular frames, 150 feet high, with a further 55 feet of tapered steelwork to form the roof areas.

Normal rolled steel girders were used, rivetted together and boxed in with steel plates.

The end walls carry the chains, from which the suspended roadways are hung, so these are stiffened with cross bracings.

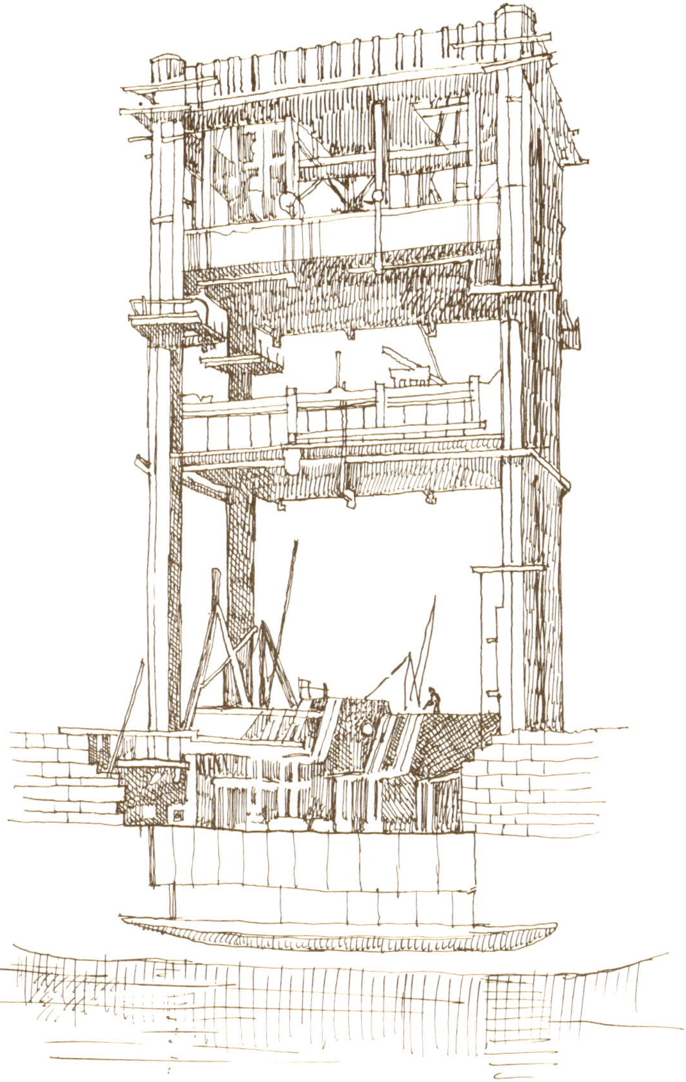

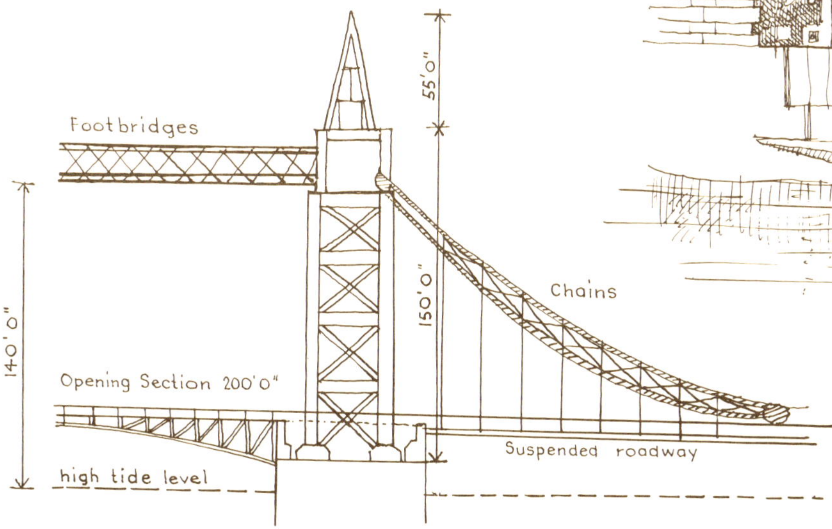

Within a month of construction work starting Sir Horace Jones died, one day after his 68th birthday.

His design intentions for the external appearance of the bridge were interpreted faithfully by another architect, George Stephenson.

As always, there was controversy about the project. Some thought the steelwork should have been exposed, others claimed that the parts that were, should have been covered up.

The anti Gothic-Revival faction did not like the ornate treatment of the stonework. Its defenders felt it blended nicely with the Tower and maintained that "civic buildings ought to look like that."

The Times hailed it as "a triumph". Other newspapers regarded it as "bald", "ugly" and a variety of other abominations.

It soon became very popular and one of London's best known features.

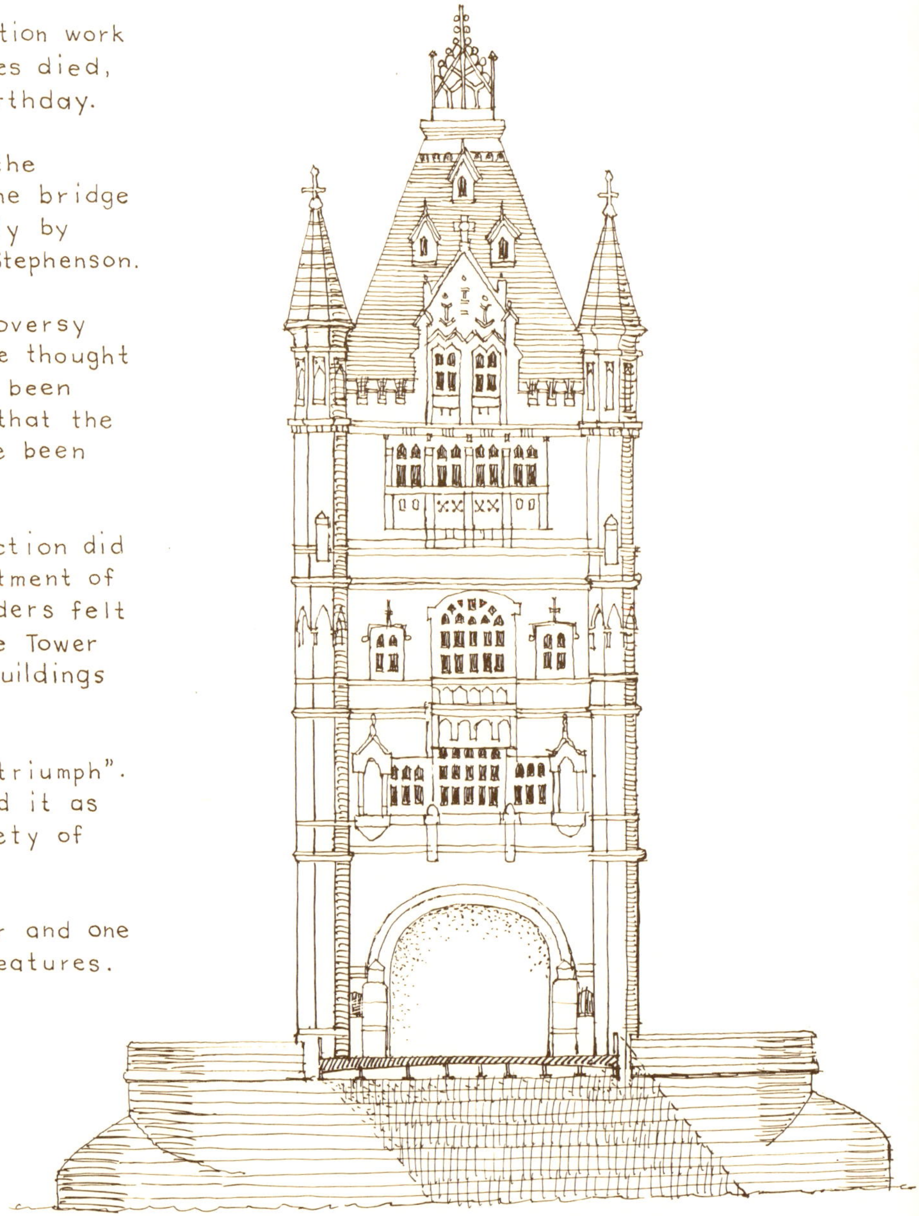

The footbridges, for which there had been such a clamour, were costly and difficult to construct. The engineer used them structurally, to stiffen the tops of the towers against the pull of the suspension chains, and each was built as a lattice box girder with the walkway inside. This enabled them to be cantilevered outwards from each tower and to be joined in the centre.

After the bridge was opened it was found that very few people used the footbridges. As sail gave way to steam the average length of wait, while the bridge was open was reduced to about six minutes and most people preferred to stand and watch the ship pass through than make the double journey up and down in the lifts and the 200 feet walk across. Eventually, Parliamentary approval was given to close the footbridges and they remained so until a few years ago.

To enable the main chains to be supported until all the links were joined, temporary wooden bridges were built between the towers and the banks of the river. Trestles were set up on these, following the line of the underside of the chains. The links were partially manufactured on shore, transported to the site, assembled and hoisted into position.

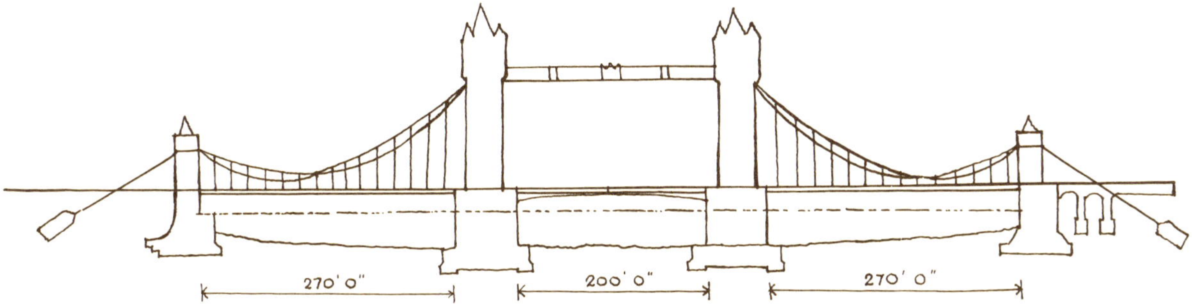

The landward ends of the chains were secured by anchor blocks of concrete, approximately 50 feet long and 30 feet wide.

From these, land ties were taken to the top of the abutment arch, at each end of the bridge, and joined to the short segment of the chain. This linked with the long segment about one third of the way along the suspension bridge.

The overall clear span of the suspension bridge on either side was 270 feet. This was then supported at about 10 feet 6 inch centres by round steel tie-rods dropping vertically from the chain.

As with the towers, the abutment arches were also steel frames cased in decorative stonework.

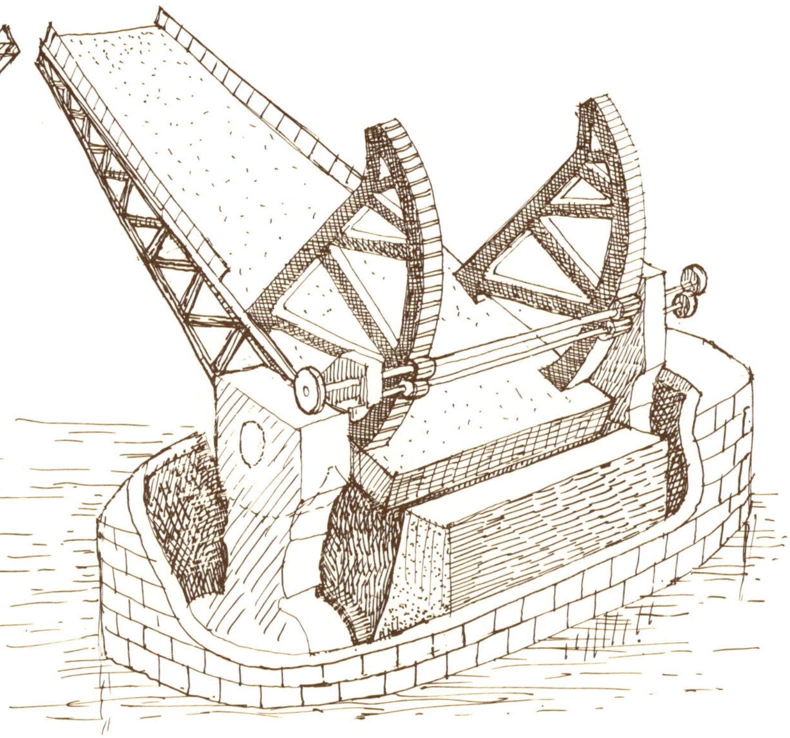

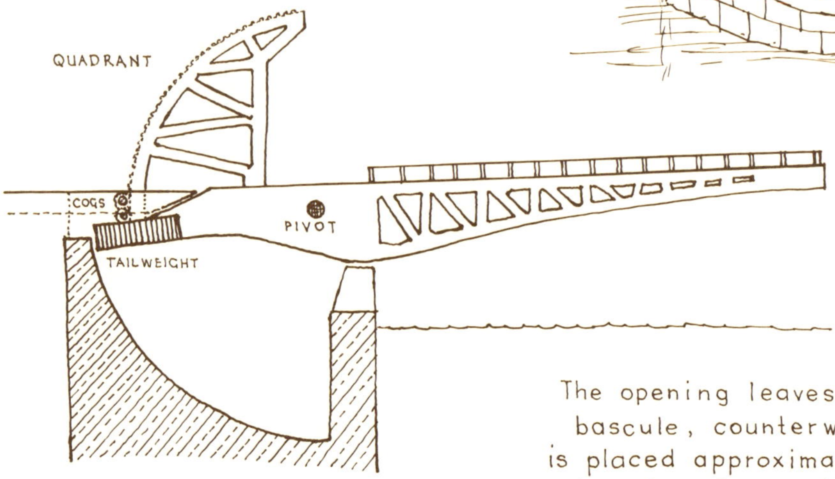

The opening leaves of the bridge are worked on the bascule, counterweighted principle. The main pivot is placed approximately a quarter of the way along the length. To equalise the weights of the two portions, 422 tons of iron and lead was built into the shorter section. This 'tailweight' greatly reduced the effort to lift the leaves. Toothed quadrants, driven by cog wheels, moved the bridge.

The total weight of each leaf is 1,045 tons carried by the main pivot.
When the bridge is down, which is most of the time, the weight is taken by the beam. Four main girders carry the moving leaves which are 100 feet long and 50 feet wide, having 32 feet of carriageway and two footpaths, each of 9 feet.

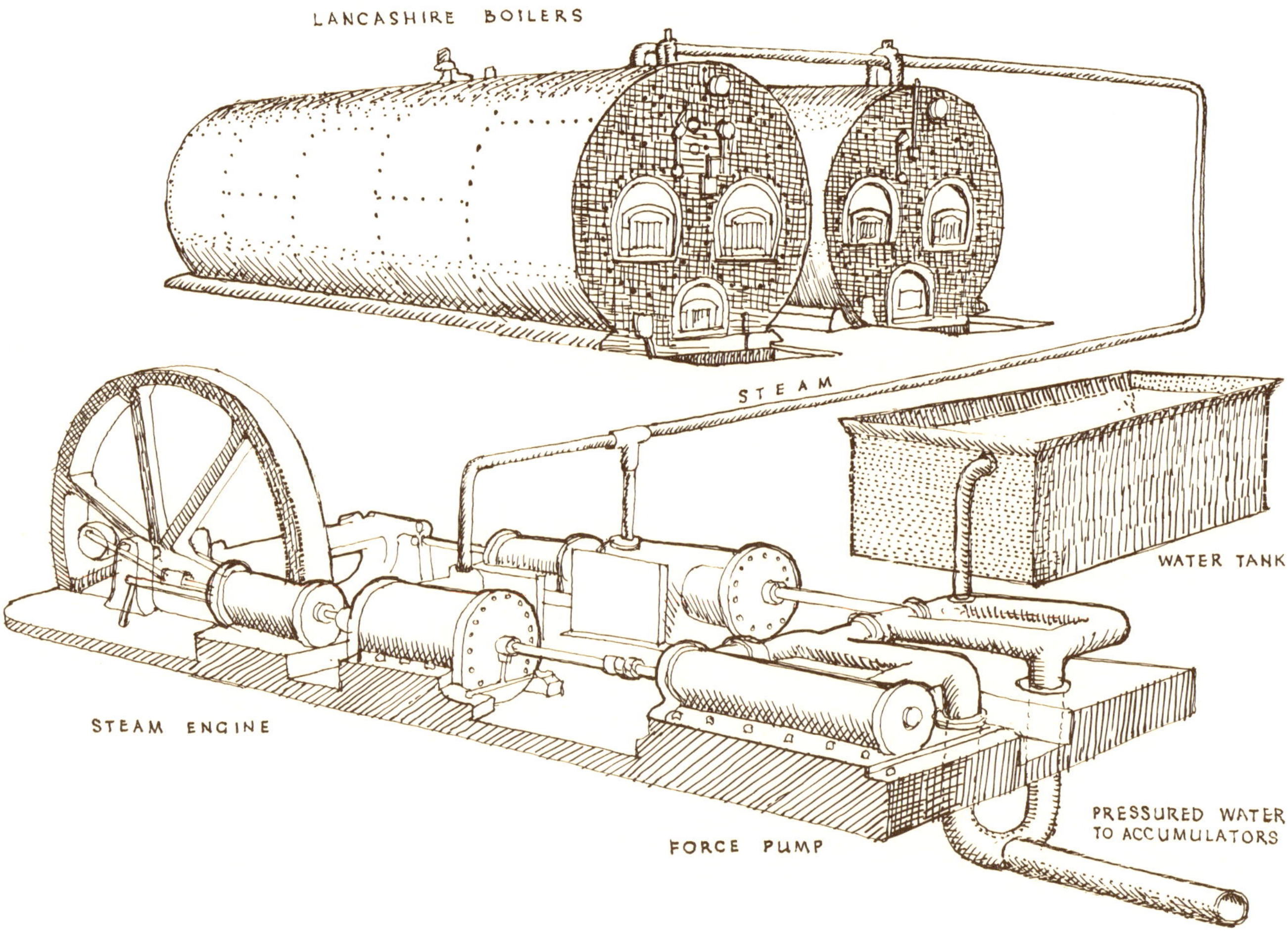

Horizontal 'Lancashire' type boilers, fired with coal, raised steam to drive the 360 hp engines.

The engines powered force pumps, which drew water from storage tanks, and it was this water under pressure that provided the motive power to operate the bascules and lifts. From the pumps the water was piped to the Main Accumulator House.

The steam engines were 'Tandem Compound' type - so called because the cylinders were paired one behind the other.

Steam from the boilers went first into the high pressure cylinders and the piston moved the flywheel. Instead of wasting the exhaust steam, it was passed into the low pressure cylinders - which were larger, because the steam had lost some of its power - and was used again to drive the engine.

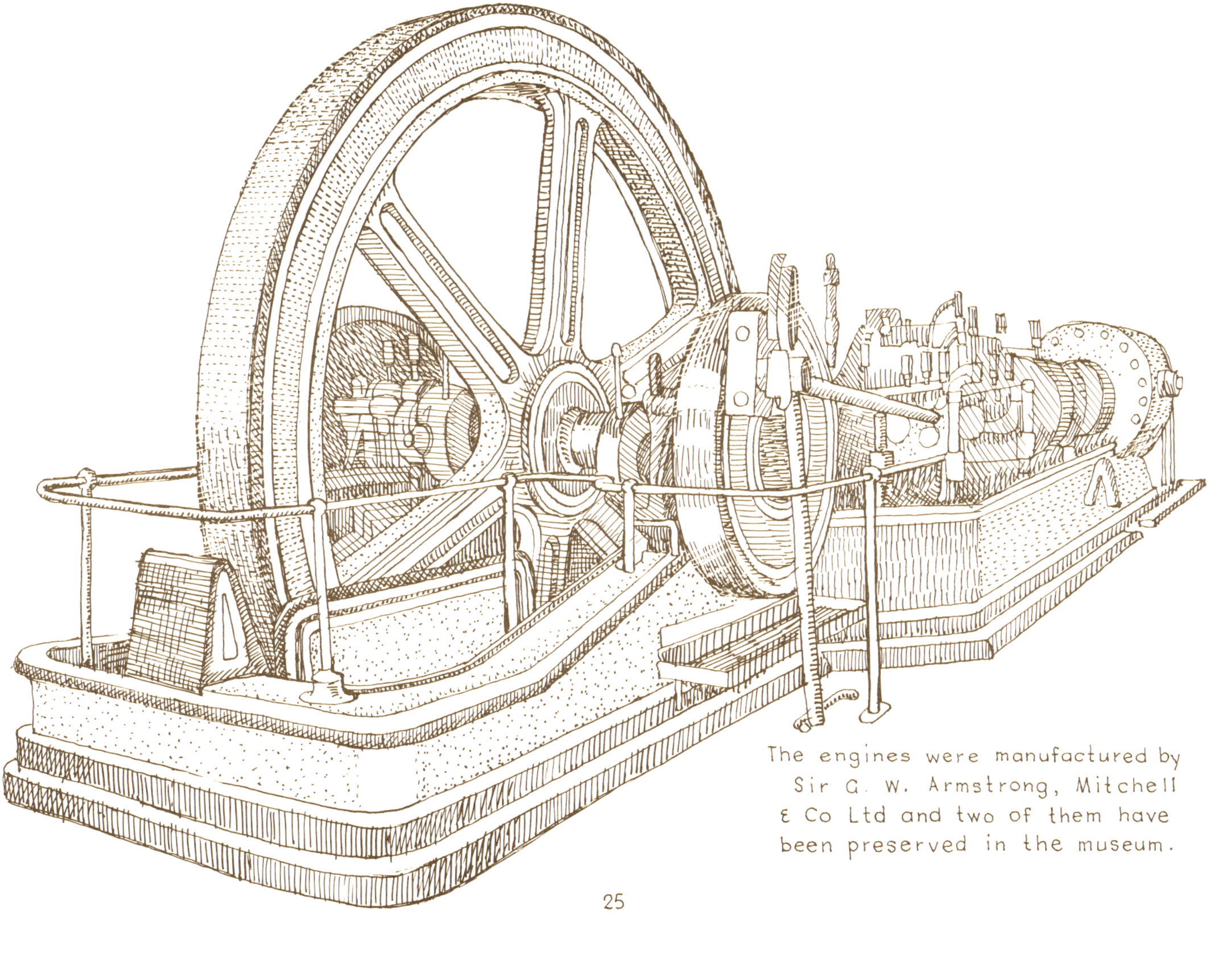

The engines were manufactured by Sir G. W. Armstrong, Mitchell & Co Ltd and two of them have been preserved in the museum.

The accumulators formed a reserve of energy always available to work the bascule engines and always being topped up by the steam driven pumps.

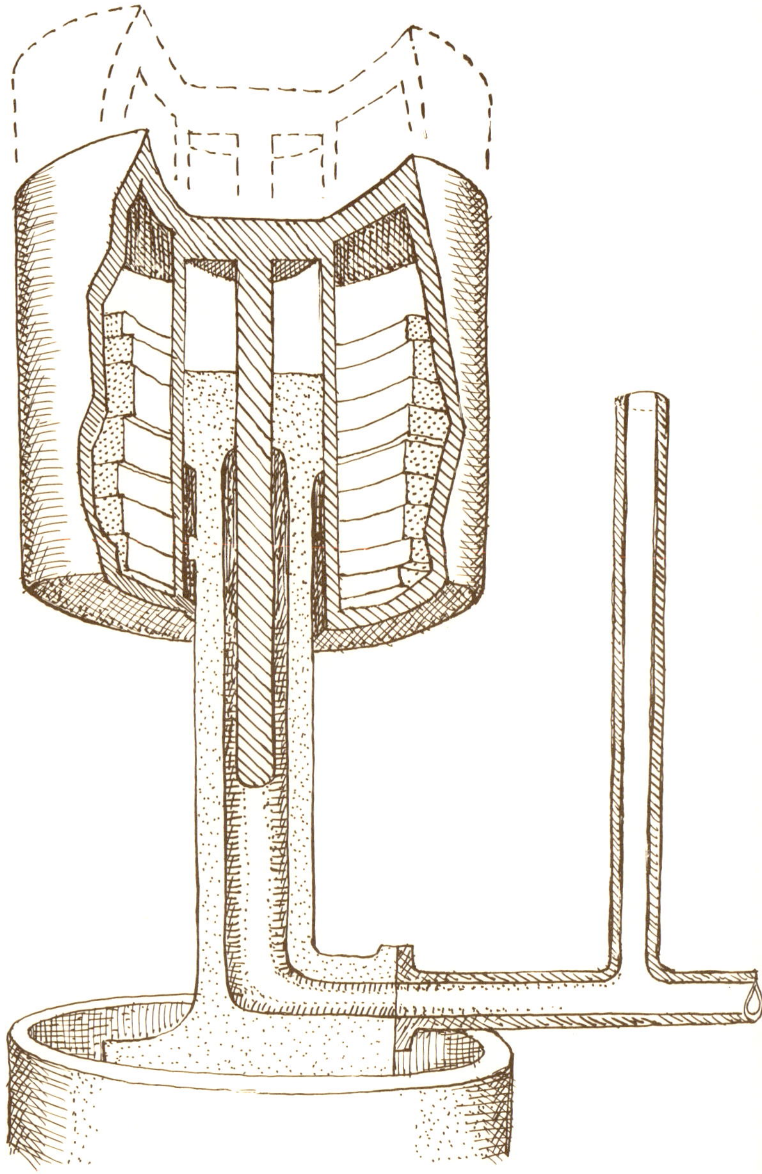

The type of accumulator used at the bridge has a central vertical cylinder approximately 35 feet high and 20 inches in diameter. Into the cylinder is fitted a piston the upper end of which is connected to an outer chamber containing weights.

As water is pumped into the cylinder the piston rises and in doing so lifts the outer chamber. The weight being about 100 tons to give a working pressure of 750 pounds per square inch.

The hydraulic engines which turned gear wheels operating the bascule drive quadrants had three single acting cylinders. When water under pressure was applied to the engine each cylinder in turn applied force to rotate the crankshaft.

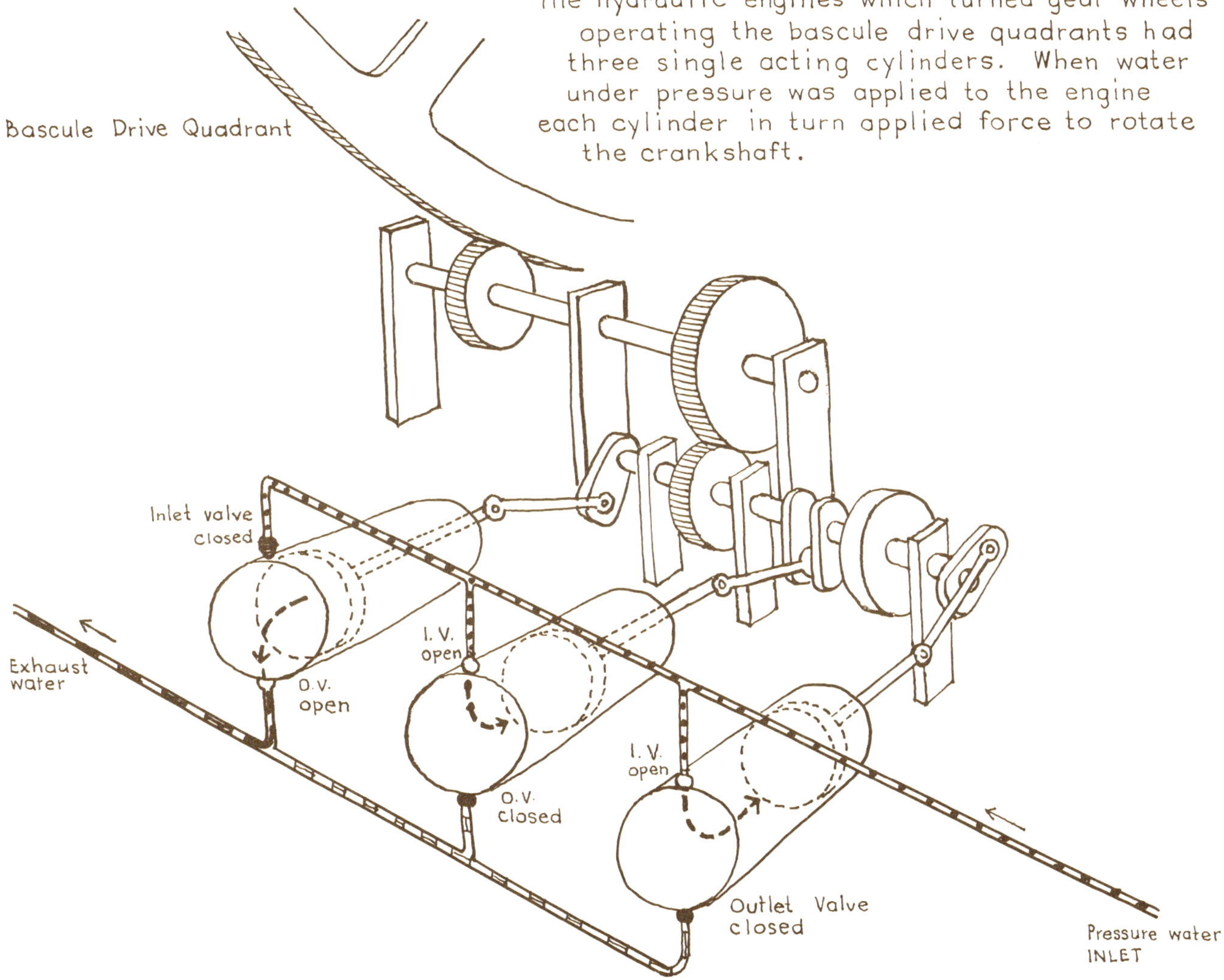

The valves controlling the flow of water into and out of the cylinders were operated automatically by the crankshaft as it rotated.

The main accumulator and engine houses were on the south bank.

Two pipes carried water across the suspension bridge to the base of the South Tower. The upper was the supply and the lower the exhaust, which returned water to the tank for re-use.

Branches from the supply fed the two accumulators there and outlets from the two hydraulic engines were connected to the return.

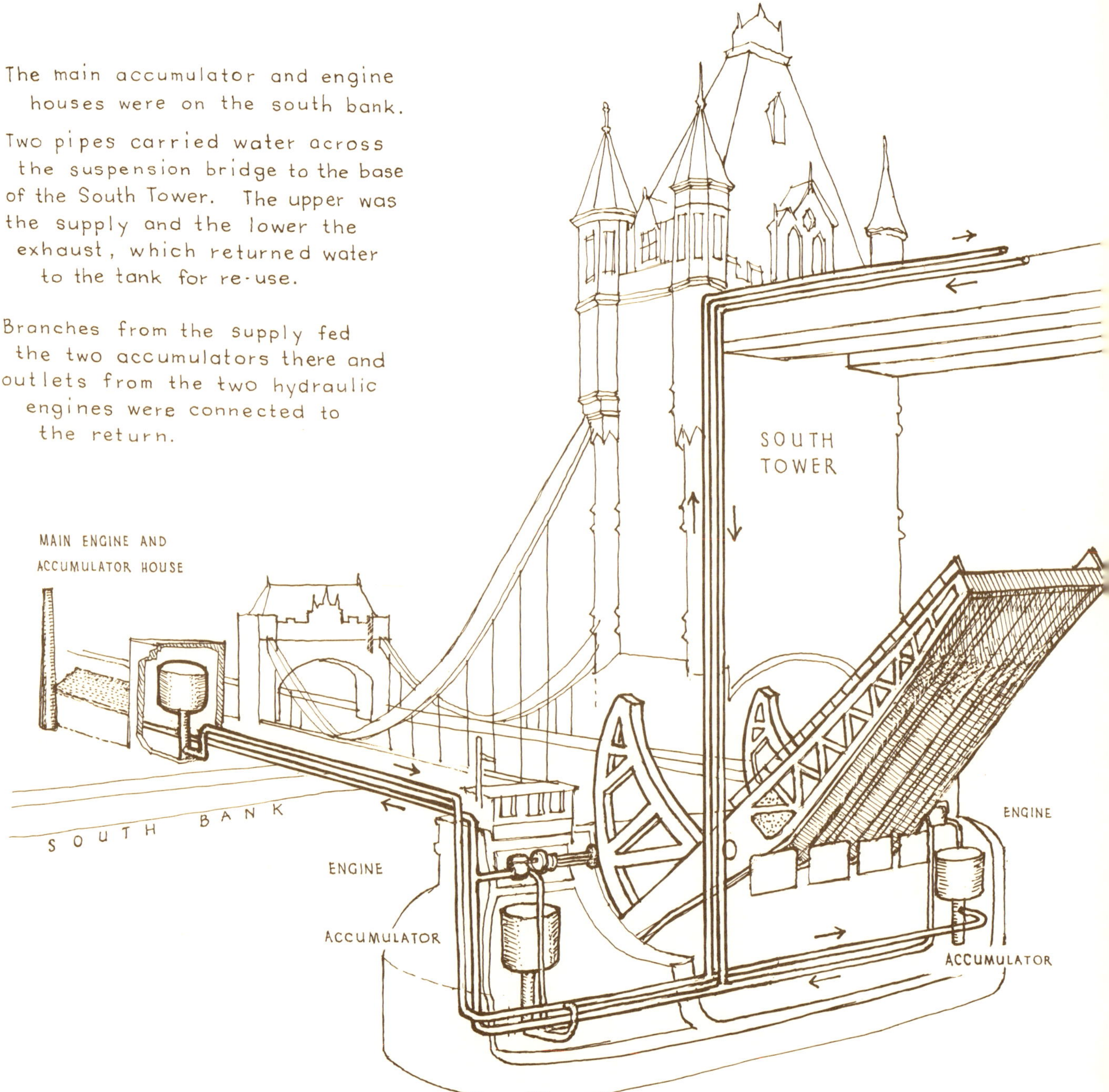

The two main pipes then continued to the North Tower, across the east footbridge, and down to the base. The other two accumulators were supplied and the exhausts of the bascule engines were connected to the return.

Passenger lifts, to give access to the footbridges, were also worked by hydraulics, powered by the same system that operated the bridge bascules.

There were two lifts in each tower, each with a capacity of up to 25 passengers. The vertical travel from road level was 110 feet and the journey took approximately one and a half minutes.

Safety was a primary concern - in a period when lift accidents were a frequent occurrence - and each lift car was fitted with six independent cables.

There were standby hydraulic chambers in case of failure.

Even though these were the very latest thing, foot travellers were still faced with two journeys in lifts, each of one and a half minutes duration, plus a walk of about 250 feet across the footbridge. Most people opted for the six minute wait below.

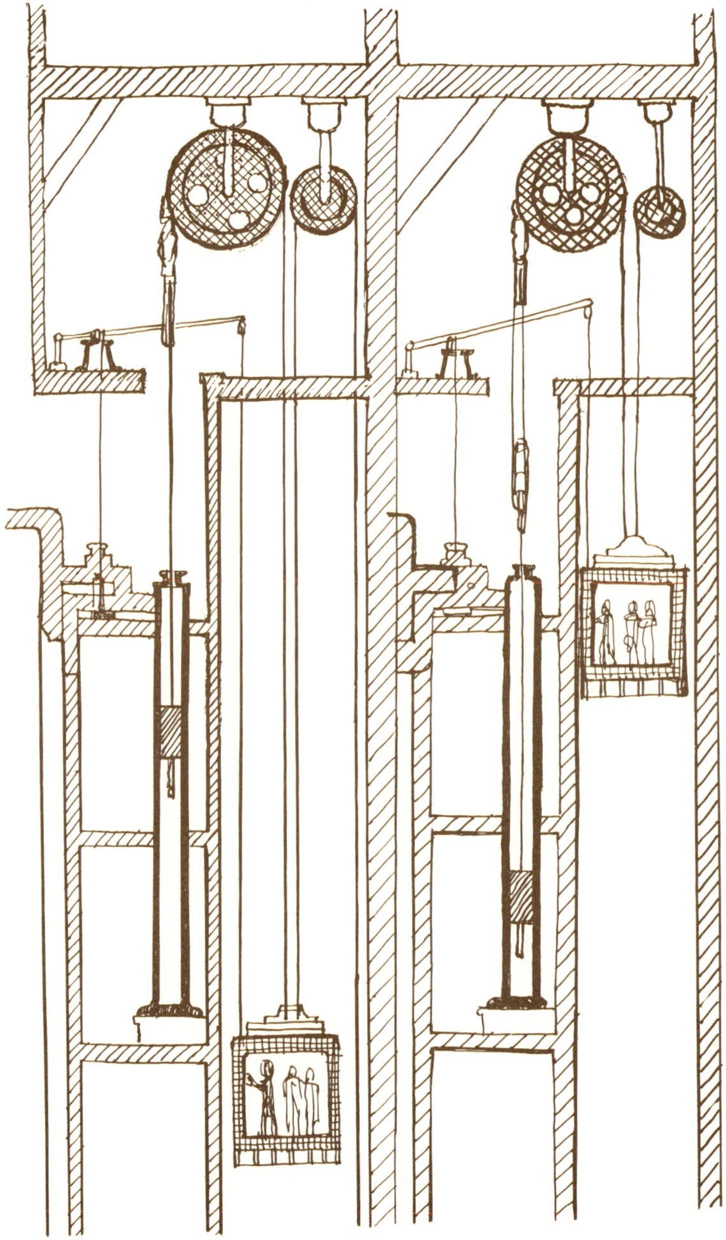

Signals were provided to show ships approaching the Bridge the position of the bascules - red when down or moving downwards, green when up or moving upwards. Semaphore was used during daylight hours and lamps after dark. In foggy conditions a gong sounded.

The machinery is controlled from cabins on the piers, in which there were originally the same kind of levers as in a railway signal box.

Shipping movements through the bridge are the responsibility of the Bridge Master. At one time, vessels could pass through at any time by simply sounding the morse code letter 'B' three times and flying a ball and pennant in daylight or two vertical red lights after dark. Now, advance notice is required before the bascules can be raised.

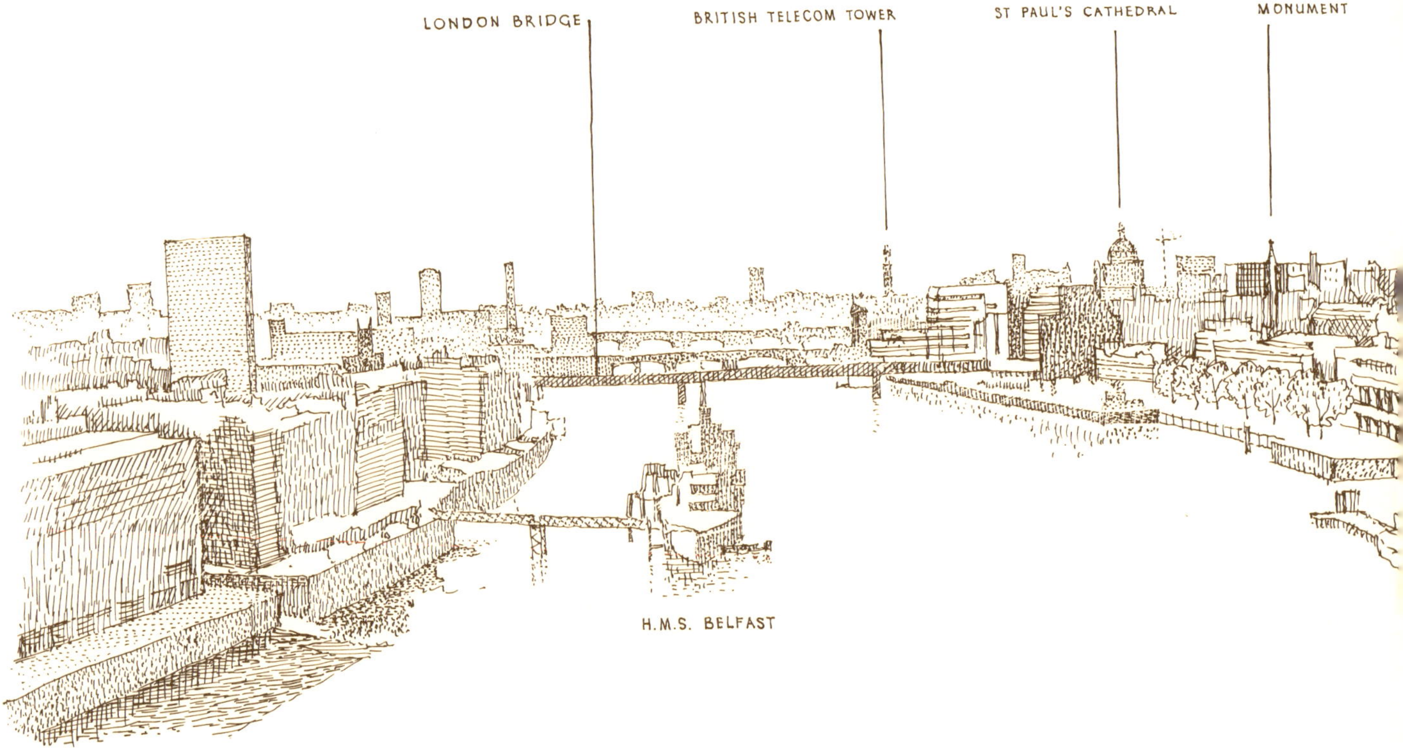

The footbridges provide magnificent panoramic views of the river and London. To the west, upstream, is the Pool of London, which until the beginning of last century was teeming with ships waiting, often for many months, to be loaded or unloaded.

London Bridge marks the western limit of the Pool and beyond are the other two bridges within the City boundary which are administered by the Bridge House Estates Committee.

H.M.S. Belfast is permanently moored off the Surrey bank as a memorial to warships of this class and is used as a naval museum.

The Tower dominates the foreground on the City shore, commanding a view over what for hundreds of years was the busiest port in the world. William the Norman established the fortress and built the White Tower. The Tudors greatly extended its boundaries and within these walls much of the darker side of English history has taken place.

Beyond is the commercial heart of the City, within which Sir Christopher Wren's St Paul's Cathedral still proudly claims its piece of skyline, linking the present with the more recent past in the capital first built by the Romans

During the first month, following its Royal Opening on 30 June 1894, Tower Bridge's centre section was raised 656 times. By the mid 1950's the total was 325,356, an average of 14 a day. The figure dropped to about 200 'lifts' in the late 70's and is now about 15 to 20 a month.

In 1982, on the same date as the original opening, the Lord Mayor of London opened the Bridge as a tourist centre and museum after a major conversion and renovation scheme had been carried out. Because Parliamentary approval had been obtained to close the footways, it was necessary to have the same approval to re-open them to the public. A large procession of ships sailed through to mark the occasion, including some that were in existence a hundred years before.

ACKNOWLEDGEMENTS:

My thanks for assistance to -

The staff of Tower Bridge.

Staff of Guildhall Library.

Staff of Royal Institute of
British Architects Library.

Staff of Bromley Libraries.

Sheila Wilson.

Published Works consulted -

"History of the Tower Bridge."
Charles Welsh FSA.
Smith, Elder & Co. 1894.

"London"
Geoffrey Trease.
Thames & Hudson, 1978.

"Mid-Victorian Masterpiece"
Barnett Cocks
Hutchinson & Co. 1977.

Contemporary copies of -
Illustrated London News.
The Times.
Current reports, guides and
news coverage.